DANGEROUS WORK!

# DIVER

RACHAEL MORLOCK



PowerKiDS press

Published in 2026 by The Rosen Publishing Group, Inc.
2544 Clinton Street, Buffalo, NY 14224



First Edition



Editor: Greg Roza
Book Design: Rachel Rising

Photo Credits: Cover, SvedOliver/Shutterstock.com; Cover, pp. 1, 3–24 Pattern image/Shutterstock.com; p. 4 Sergiy Zavgorodny/Shutterstock.com; p. 5 troycini/Shutterstock.com; p. 6 Vangelis_Vassalakis/Shutterstock.com; p. 7 Arseniy Bobrov/Shutterstock.com; p.9 Nomad_Soul/Shutterstock.com; p. 11 Calypso Photo/Shutterstock.com; p. 13 Sakis Lazarides/Shutterstock.com; p. 14 Benny Marty/Shutterstock.com; p. 15 zaferkizilkaya/Shutterstock.com; p. 17 SvedOliver/Shutterstock.com; p. 18 OlegDoroshin/Shutterstock.com; p. 19 Roxane 134/Shutterstock.com; p. 20 chonlasub woravichan/Shutterstock.com; p. 21 blue-sea.cz/Shutterstock.com.

Library of Congress Cataloging-in-Publication Data

Names: Morlock, Rachael, author.
Title: Diver / Rachael Morlock.
Description: [Buffalo, New York] : PowerKids Press, [2026] | Series: Dangerous work! | Includes bibliographical references and index.
Identifiers: LCCN 2025412016 (print) | LCCN 2025412017 (ebook) | ISBN 9781499452341 (library binding) | ISBN 9781499452334 (paperback) | ISBN 9781499452358 (ebook)
Subjects: LCSH: Deep diving–Vocational guidance–Juvenile literature. | Deep diving–Safety measures–Juvenile literature.
Classification: LCC VM984 .M67 2026 (print) | LCC VM984 (ebook) | DDC 627/.72–dc23/eng/20250212
LC record available at https://lccn.loc.gov/2025412016
LC ebook record available at https://lccn.loc.gov/2025412017

Manufactured in the United States of America



CPSIA Compliance Information: Batch #CSPK26. For Further Information contact Rosen Publishing at 1-800-237-9932.

# CONTENTS

# UNDERWATER WORK

It takes strength and skill to be an underwater diver. A lot of people learn just enough about diving to do it for fun. Imagine a vacation spent in warm waters, diving deep enough to see colorful coral reefs or friendly **marine** animals!

Other people turn their love for diving into a job. These working divers combine their knowledge of diving with other special skills. They might study sea life, fix bridges, or search for missing people, all while underwater!

People have practiced underwater diving since ancient times. Diving for pearls or sponges are very old jobs. In this photo, a diver farms oysters to produce pearls.

# GETTING DEEPER

Humans have probably been diving for as long as they've been swimming. However, it's taken hundreds of years and the ideas of many creative and curious inventors to make diving what it is today.

Most diving jobs require people to spend a long time underwater. The tools that make those jobs possible are fairly new. A great breakthrough came in 1943, when Jacques-Yves Cousteau and Emile Gagnon created the Aqua-Lung—a breathing **apparatus** that made diving easier and safer.

## STAY SAFE!

**How do divers breathe underwater? With an underwater breathing apparatus! The apparatus could either use tanks of gas carried by the diver, or a kind of tube that supplies breathing gases from the surface of the water down to the diver.**

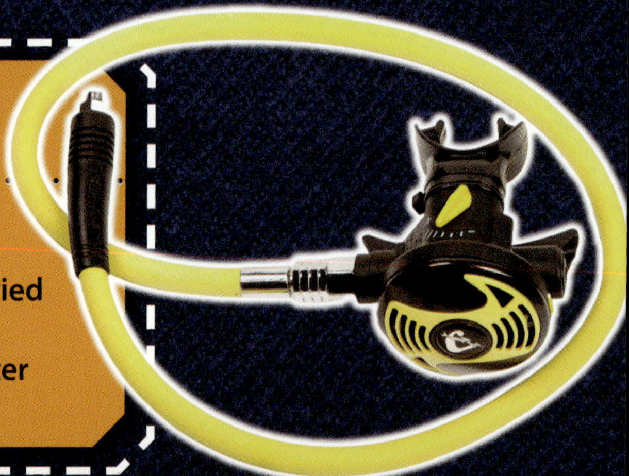

This diver carries an air supply in tanks. Divers have to regularly check their tanks while they're underwater to make sure they have enough air.

# BE PREPARED

Underwater diving is a dangerous business. Whether you're diving for fun or for work, you must know how to use gear properly and follow safety rules. For that, training is necessary.

Working divers go beyond basic dive training. They earn **certifications** by taking special courses and logging a certain number of dives. All working divers need a Divemaster certification. Divemasters are trained in open water diving, first aid and **CPR**, and rescue dives. The best divers are also knowledgeable about the science of diving.

## STAY SAFE!

**If you want to be a diver, you can start by learning to swim. Diving and swimming are separate skills—you use your legs more for diving and your arms more for swimming. Still, swim practice will make you more comfortable and safe in the water.**

A number of organizations around the world offer dive training. Students practice in a swimming pool before testing their skills in open water.

# THE SCIENCE OF DIVING

The more you know about the science of diving, the safer you are in the water. That's why working divers spend time studying key scientific ideas. They have to be aware of the way **pressure**, currents, tides, heat, light, buoyancy, and more can affect a dive.

A diver will either sink or float in the water, based on their buoyancy. Divers must balance the forces of buoyancy (which pushes them up) and gravity (which pushes them down) to control their movements underwater.

## STAY SAFE!

If a diver is in cold water, their body will quickly lose heat. This can cause the condition **hypothermia**. Wearing a wetsuit or drysuit helps keep a diver's body **temperature** in the safe zone.

This diver uses a weight belt as a tool to help them control buoyancy.

# UNDER PRESSURE

Dealing with changes in pressure is one of a diver's most important tasks. The deeper the dive, the greater the pressure in the water. In high pressure, a gas called nitrogen builds up in the body.

**Recreational** divers go less than 130 feet (40 m) below the surface of the water. However, working divers can reach 300 feet (90 m)! They must come up from deep or long dives slowly. Their bodies need time to get rid of extra nitrogen and return to lower pressure.

## STAY SAFE!

If divers come up from a deep dive too quickly, nitrogen bubbles form in the body. This is called decompression sickness. It's also called "the bends" because it can make people unable to straighten their **joints**. The condition is painful and dangerous.

Divers make decompression stops as they return to the surface from a deep dive. They use tools called decompression tables and dive computers to time and plan each stop.

# SCIENTIFIC DIVING

Many working divers are scientists. They're masters of diving, but they're also trained in a scientific field. Marine biologists use dives to learn about sea life. They might observe sea creatures, look for plants, or take measurements and **samples** from underwater.

Marine archaeologists learn about human history by searching the water for objects made by people. For example, they might dive to a shipwreck. The things they find can tell them more about what happened, or about how people lived in the past.

Not every marine biologist is a diver, but being able to dive is a helpful skill! Diving brings these scientists closer to the world they study.

# COMMERCIAL DIVING

Commercial divers combine their diving skills with another trade. There are two groups of commercial diving jobs—offshore and inland. Offshore workers perform dives in the ocean. Many work on big ships that carry oil. Usually these divers do special jobs underwater to repair, clean, or **inspect** the ship.

Inland commercial divers work in smaller bodies of water like rivers, lakes, or canals. Many inland divers repair or inspect bridges, dams, ferries, power plants, or other structures built in water.

## STAY SAFE!

**Mixing electricity and water is very dangerous, but that's what a lot of commercial divers have to do! Their work is underwater welding, which uses electricity to join pieces of metal together. Special tools and training help divers do this task safely.**

Divers learn how to weld on dry land before taking their skills underwater. Underwater welders fix ships, bridges, pipelines, dams, and more.

# PUBLIC SAFETY

Many members of the military and police force are skilled divers. They're often trained to do search and rescue work in the case of an accident or **emergency**. They look for missing people who have been lost or hurt in the water.

Police divers are also trained to search for clues underwater after a crime. Military divers have different jobs depending on their branch of the military. Some learn how to find bombs underwater and stop them from exploding.

Search and rescue work can be upsetting and stressful. It's important for divers to stay calm and think clearly while they're in the water.

# DIVE RIGHT IN!

People have found dozens of ways to combine diving with other jobs. For example, some divers work in the entertainment business. They make sure actors and crew members are safe when they film underwater scenes for movies or television. Other divers are teachers. They share their love of the sport with their students and prepare them for safe dives.

Diving may be dangerous, but it's also a lot of fun. With good training and smart planning, diving can turn everyday work into an adventure!

Underwater photographers use their dives to capture photos and videos of the plants, animals, and wonders of the blue world.

# GLOSSARY

**apparatus:** The tools or objects needed to perform a task.

**certification:** Official approval that someone has been trained to do something.

**CPR:** Also called cardiopulmonary resuscitation, a method of restarting a person's heartbeat and breathing.

**emergency:** An unexpected situation that needs quick action.

**hypothermia:** Dangerously low body temperature caused by cold conditions.

**inspect:** To look at something carefully in order to learn more about it or find problems with it.

**joint:** A point where two bones meet in the body.

**marine:** Having to do with the sea.

**pressure:** A force that pushes on something else.

**recreational:** Something done for pleasure or relaxation.

**sample:** A small part of something.

**temperature:** How hot or cold something is.

# FOR MORE INFORMATION

## BOOKS

Noelle, Becky. *Scuba Diving*. New York, NY: AV2, 2022.

Petrie, Kristin. *Jacques Cousteau*. Minneapolis, MN: Checkerboard Library, 2022.

## WEBSITES

**Kiddle: Scuba Diving**
*kids.kiddle.co/Scuba_diving*
Read more about the history of scuba diving and the inventions that make it possible.

**Ology: Marine Biology**
*www.amnh.org/explore/ology/marine-biology*
Learn about some of the topics marine biologists study!

Publisher's note to educators and parents: Our editors have carefully reviewed these websites to ensure that they are suitable for students. Many websites change frequently, however, and we cannot guarantee that a site's future contents will continue to meet our high standards of quality and educational value. Be advised that students should be closely supervised whenever they access the internet.

# INDEX